MONEY
WORKBOOK

SHOPPING PROBLEMS

hot dog = $1.00	cola = $1.00
order of French-fries = $1.00	ice cream cone = $1.50
hamburger = $2.00	milk shake = $2.00
deluxe cheeseburger = $3.00	taco = $2.50

1. .... What is the total cost of five milk shakes, three hot dogs, two ice cream cones, and five tacos?

2. .... Brian wants to buy an order of French-fries and four ice cream cones. How much will he have to pay?

3. .... If Adam wanted to buy three milk shakes, five ice cream cones, and four hamburgers, how much money would he need?

4. .... What is the total cost of two hamburgers, two hot dogs, a cola, and five milk shakes?

5. .... If Donald wanted to buy four hamburgers, an order of French-fries, four ice cream cones, and a deluxe cheeseburger, how much money would he need?

6. .... If Michele buys two colas, what will her change be if she pays $10.00?

7. .... What is the total cost of five orders of French-fries, three hot dogs, five milk shakes, and three tacos?

8. .... Steven purchases five milk shakes, two hamburgers, and three ice cream cones. How much money will he get back if he pays $25.00?

hot dog = $1.00	cola = $1.00
order of French-fries = $1.00	ice cream cone = $1.50
hamburger = $2.50	milk shake = $2.50
deluxe cheeseburger = $3.00	taco = $2.00

1. .... What is the total cost of four colas?

2. .... Jake wants to buy a cola, five ice cream cones, and five deluxe cheeseburgers. How much will he have to pay?

3. .... What is the total cost of three hamburgers?

4. .... If Billy wanted to buy three hamburgers, five deluxe cheeseburgers, a milk shake, and an order of French-fries, how much would he have to pay?

5. .... What is the total cost of four orders of French-fries and four ice cream cones?

6. .... If Marin wanted to buy a hot dog, three hamburgers, two deluxe cheeseburgers, and five orders of French-fries, how much would she have to pay?

7. .... Sharon wants to buy two tacos. How much will it cost her?

8. .... Jennifer purchases three tacos and four milk shakes. If she had $20.00, how much money will she have left?

hot dog = $1.50	cola = $1.00
order of French-fries = $0.50	ice cream cone = $1.50
hamburger = $2.00	milk shake = $2.00
deluxe cheeseburger = $3.50	taco = $2.00

1. .... Marcie wants to buy four milk shakes, two colas, and five ice cream cones. How much will it cost her?

2. .... David purchases five milk shakes. How much money will he get back if he pays $20.00?

3. .... What is the total cost of four orders of French-fries, four milk shakes, two hot dogs, a taco, and a deluxe cheeseburger?

4. .... If Amy buys two deluxe cheeseburgers, an order of French-fries, two hot dogs, and three tacos, what will her change be if she pays $30.00?

5. .... Sandra purchases four orders of French-fries. If she had $10.00, how much money will she have left?

6. .... Jennifer purchases four milk shakes, three orders of French-fries, two hamburgers, and five ice cream cones. How much change will she get back from $25.00?

7. .... What is the total cost of four orders of French-fries and three milk shakes?

8. .... If Adam wanted to buy a deluxe cheeseburger, two tacos, an ice cream cone, and two orders of French-fries, how much would he have to pay?

hot dog = $1.50	cola = $1.00
order of French-fries = $1.00	ice cream cone = $1.00
hamburger = $2.50	milk shake = $2.00
deluxe cheeseburger = $3.00	taco = $2.50

1. .... If Jennifer buys three ice cream cones, what will her change be if she pays $10.00?

2. .... What is the total cost of five orders of French-fries, three hot dogs, three hamburgers, and a milk shake?

3. .... If Ellen buys four deluxe cheeseburgers, two tacos, four ice cream cones, a hot dog, and a milk shake, and if she had $30.00, how much money will she have left?

4. .... Audrey purchases three milk shakes. What will her change be if she pays $20.00?

5. .... If Adam buys three deluxe cheeseburgers, two milk shakes, three hamburgers, five tacos, and four orders of French-fries, and if he had $50.00, how much money will he have left?

6. .... What is the total cost of four ice cream cones?

7. .... If Jake buys four hot dogs and a hamburger, how much money will he get back if he pays $20.00?

8. .... Jackie wants to buy five milk shakes, four deluxe cheeseburgers, three colas, a hot dog, and a hamburger. How much will it cost her?

hot dog = $1.00	cola = $1.00
order of French-fries = $0.50	ice cream cone = $1.00
hamburger = $2.50	milk shake = $2.00
deluxe cheeseburger = $3.50	taco = $2.00

1. .... If Billy wanted to buy an order of French-fries, four ice cream cones, and four tacos, how much would he have to pay?

2. .... Amy wants to buy five milk shakes. How much will she have to pay?

3. .... Brian purchases four ice cream cones and a hamburger. How much money will he get back if he pays $20.00?

4. .... Allan wants to buy two milk shakes, five tacos, three hamburgers, a deluxe cheeseburger, and a hot dog. How much will it cost him?

5. .... Jackie wants to buy five hot dogs and two tacos. How much will it cost her?

6. .... Ellen purchases a deluxe cheeseburger and five ice cream cones. How much change will she get back from $20.00?

7. .... If Donald buys five tacos, four hamburgers, and three ice cream cones, what will his change be if he pays $30.00?

8. .... Steven purchases three orders of French-fries. How much money will he get back if he pays $10.00?

hot dog = $1.00	cola = $1.00
order of French-fries = $1.00	ice cream cone = $1.50
hamburger = $2.50	milk shake = $2.50
deluxe cheeseburger = $3.50	taco = $2.00

1. .... Michele purchases three milk shakes, two orders of French-fries, and four colas. If she had $20.00, how much money will she have left?

2. .... Allan wants to buy three tacos. How much will it cost him?

3. .... What is the total cost of a hot dog?

4. .... If Jake buys five orders of French-fries, how much change will he get back from $10.00?

5. .... Amy wants to buy a milk shake, a cola, an order of French-fries, five ice cream cones, and four hot dogs. How much will she have to pay?

6. .... If Sandra buys three colas and five tacos, what will her change be if she pays $20.00?

7. .... What is the total cost of a hot dog and four tacos?

8. .... If Jennifer buys three hamburgers, four tacos, and three orders of French-fries, how much money will she get back if she pays $30.00?

hot dog = $1.50	cola = $1.00
order of French-fries = $0.50	ice cream cone = $1.00
hamburger = $2.00	milk shake = $2.00
deluxe cheeseburger = $3.50	taco = $2.50

1. .... What is the total cost of a milk shake, five tacos, and two deluxe cheeseburgers?

2. .... Ellen wants to buy five ice cream cones, an order of French-fries, and two hamburgers. How much money will she need?

3. .... If Audrey wanted to buy five tacos, four deluxe cheeseburgers, two colas, and four milk shakes, how much money would she need?

4. .... If Michele buys three deluxe cheeseburgers and two hamburgers, how much money will she get back if she pays $20.00?

5. .... If Sandra wanted to buy two hamburgers and three deluxe cheeseburgers, how much money would she need?

6. .... If Donald wanted to buy three orders of French-fries, a hot dog, three tacos, two ice cream cones, and four colas, how much would it cost him?

7. .... If Sharon wanted to buy an ice cream cone, four deluxe cheeseburgers, three orders of French-fries, and two hot dogs, how much money would she need?

8. .... Billy purchases four tacos, three hamburgers, two orders of French-fries, two hot dogs, and three colas. How much change will he get back from $30.00?

hot dog = $1.50	cola = $1.00
order of French-fries = $1.00	ice cream cone = $1.00
hamburger = $2.00	milk shake = $2.00
deluxe cheeseburger = $3.00	taco = $2.00

1. .... What is the total cost of five deluxe cheeseburgers, five tacos, three hamburgers, and five colas?

2. .... Adam purchases two colas. What will his change be if he pays $5.00?

3. .... What is the total cost of a hamburger and a hot dog?

4. .... Sandra wants to buy a milk shake, five deluxe cheeseburgers, an ice cream cone, three hot dogs, and two tacos. How much will it cost her?

5. .... Audrey wants to buy four ice cream cones and five colas. How much money will she need?

6. .... If Sharon buys an ice cream cone and four deluxe cheeseburgers, how much change will she get back from $20.00?

7. .... Jackie wants to buy a deluxe cheeseburger. How much will she have to pay?

8. .... David purchases three tacos and five deluxe cheeseburgers. What will his change be if he pays $30.00?

hot dog = $1.00	cola = $1.00
order of French-fries = $0.50	ice cream cone = $1.50
hamburger = $2.50	milk shake = $2.50
deluxe cheeseburger = $3.50	taco = $2.50

1. .... Steven wants to buy two colas, two deluxe cheeseburgers, a hot dog, and five milk shakes. How much will he have to pay?

2. .... Janet wants to buy five hot dogs, three hamburgers, and three tacos. How much money will she need?

3. .... If David buys three colas, three hamburgers, two orders of French-fries, two milk shakes, and a taco, how much money will he get back if he pays $30.00?

4. .... Sharon purchases five orders of French-fries, an ice cream cone, and five tacos. How much money will she get back if she pays $30.00?

5. .... Marin wants to buy two ice cream cones, two colas, and five hot dogs. How much will she have to pay?

6. .... If Donald wanted to buy two tacos, how much would he have to pay?

7. .... If Jackie wanted to buy a taco, two hamburgers, and an order of French-fries, how much would she have to pay?

8. .... If Ellen buys four colas, a hot dog, four milk shakes, five orders of French-fries, and four hamburgers, what will her change be if she pays $40.00?

hot dog = $1.00	cola = $1.00
order of French-fries = $1.00	ice cream cone = $1.50
hamburger = $2.00	milk shake = $2.50
deluxe cheeseburger = $3.00	taco = $2.50

1. .... If David buys two tacos and two deluxe cheeseburgers, and if he had $15.00, how much money will he have left?

2. .... If Donald wanted to buy two ice cream cones, two hot dogs, four tacos, three milk shakes, and two hamburgers, how much money would he need?

3. .... If Allan wanted to buy two tacos, four orders of French-fries, three ice cream cones, and three deluxe cheeseburgers, how much money would he need?

4. .... What is the total cost of a hamburger?

5. .... Audrey wants to buy a hamburger, two ice cream cones, five orders of French-fries, three deluxe cheeseburgers, and a taco. How much will she have to pay?

6. .... Sandra wants to buy two milk shakes. How much will it cost her?

7. .... Janet purchases three hamburgers. How much change will she get back from $20.00?

8. .... Jake purchases two hamburgers, five tacos, five deluxe cheeseburgers, three hot dogs, and three ice cream cones. If he had $50.00, how much money will he have left?

COUNTING MONEY

Emma.School

1.   = ...

    ____ ____ ____ ____

2.    = ...

    ____ ____ ____ ____

3.   = ...

    ____ ____ ____

4.    = ...

    ____ ____ ____ ____

5.     = ...

    ____ ____ ____ ____ ____

6.   = ...

    ____ ____ ____

7.    = ...

    ____ ____ ____ ____

8.    = ...

    ____ ____ ____ ____ ____

1.    = ...

_____ _____ _____ _____ _____

2.    = ...

_____ _____ _____

3.   = ...

_____ _____ _____

4.      = ...

_____ _____ _____ _____ _____

5.     = ...

_____ _____ _____ _____

6.   = ...

_____ _____

7.   = ...

_____ _____ _____

8.    = ...

_____ _____ _____ _____

_____

1.

____   ____   ____   = ...

2.

____   ____   ____   ____   = ...

3.

____   ____   ____   ____   = ...

____

4.

____   ____   ____   ____   = ...

5.

____   ____   ____   = ...

6.

____   ____   ____   = ...

7.

____   ____   ____   ____   = ...

8.

____   ____   ____   ____   = ...

____

1.   = ...

    ___   ___   ___

2.     = ...

    ___   ___   ___   ___   ___

3.    = ...

    ___   ___   ___   ___

4.    = ...

    ___   ___   ___   ___

    ___

5.    = ...

    ___   ___   ___

6.    = ...

    ___   ___   ___   ___

7.     = ...

    ___   ___   ___   ___

8. = ...

    ___   ___   ___

1.   = ...

____ ____ ____ _____

2.   = ...

____ _____ _____ _____ _____

3. = ...

____ _____ _____

4.   = ...

____ _____ _____ _____

5.   = ...

____ _____ _____

6.    = ...

_____ _____ _____ _____

7.   = ...

_____ _____ _____

8.    = ...

____ _____ _____ _____

Emma.Schoo

1.   = ...

___   _____   _____

2.    = ...

_____   _____   _____   _____

3.    = ...

_____   _____   _____   _____

4.    = ...

___   _____   _____   _____

5.   = ...

___   _____   _____

6.      = ...

___   _____   _____   _____   _____

7.    = ...

___   _____   _____

8.     = ...

___   _____   _____   _____   _____

1.  = ...

____ ____ ____ _____

2.    = ...

____ ____ _____

3.    = ...

_____ _____ _____ _____

_____

4.     = ...

____ _____ _____ _____

5.    = ...

____ _____ _____ _____

6.    = ...

_____ _____ _____

7.     = ...

____ ____ _____ _____

8.    = ...

_____ _____ _____

1.   = ...

———— ———— ————   ————

2.    = ...

———— ———— ————   ————

3.   = ...

———— ———— ————

4.   = ...

———— ———— ————

5.    = ...

———— ———— ———— ———— ————

6.     = ...

———— ———— ———— ————

7.   = ...

———— ———— ————

8.     = ...

———— ———— ———— ————

————

Emma.School

1.    = ...

___ ___ _____ _____ _____

2.    = ...

_____ _____ _____

3.   = ...

_____ _____ _____

4.     = ...

_____ _____ _____

_____

5.    = ...

_____ _____ _____

6.    = ...

___ _____ _____

7.     = ...

_____ _____ _____ _____

8.      = ...

___ _____ _____ _____ _____

1.    = ...

___   ___   _____   _____   _____

2.   = ...

___   ___   _____   _____

3.   = ...

___   ___   _____

4.    = ...

___   ___   _____   _____

5.   = ...

___   ___   _____

6.     = ...

___   ___   _____   _____

7.     = ...

___   ___   _____   _____

8.   = ...

___   ___   _____

WAGES

1. How much will David earn if he earns $4.00 per hour and works 23 hours?

2. If Marin earns $308.00 after working 44 hours what is the hourly rate?

3. Jennifer baby-sat for 30 hours over two weeks. She earned $9.00 an hour. What was her gross pay?

4. How much will Donald earn if he earns $4.00 per hour and works 48 hours?

5. If Ellen earns $225.00 after working 25 hours what is the hourly rate?

1. How much will Jake earn if he earns $7.00 per hour and works 32 hours?

   ....

2. If Michele earns $91.00 after working 13 hours what is the hourly rate?

   ....

3. Janet baby-sat for 49 hours over two weeks. She earned $4.00 an hour. What was her gross pay?

   ....

4. If Jennifer earns $82.00 after working 41 hours what is the hourly rate?

   ....

5. Marcie baby-sat for eight hours over two weeks. She earned $1.00 an hour. What was her gross pay?

   ....

1. If Marin earns $238.00 after working 34 hours what is the hourly rate?

   ....

2. How much will Adam earn if he earns $7.00 per hour and works 43 hours?

   ....

3. Ellen baby-sat for 24 hours over two weeks. She earned $3.00 an hour. What was her gross pay?

   ....

4. How much will Donald earn if he earns $3.00 per hour and works eight hours?

   ....

5. Jackie baby-sat for 47 hours over two weeks. She earned $1.00 an hour. What was her gross pay?

   ....

1. How much will Donald earn if he earns $4.00 per hour and works 23 hours?

....

2. Ellen baby-sat for 38 hours over two weeks. She earned $3.00 an hour. What was her gross pay?

....

3. If Jennifer earns $108.00 after working 36 hours what is the hourly rate?

....

4. Janet baby-sat for 36 hours over two weeks. She earned $9.00 an hour. What was her gross pay?

....

5. How much will David earn if he earns $9.00 per hour and works 44 hours?

....

1. If Janet earns $234.00 after working 39 hours what is the hourly rate?

2. Jennifer baby-sat for 28 hours over two weeks. She earned $9.00 an hour. What was her gross pay?

3. How much will Brian earn if he earns $3.00 per hour and works 38 hours?

4. If Marcie earns $11.00 after working 11 hours what is the hourly rate?

5. Audrey baby-sat for 25 hours over two weeks. She earned $3.00 an hour. What was her gross pay?

1. If Jennifer earns $45.00 after working 45 hours what is the hourly rate?

....

2. How much will Allan earn if he earns $3.00 per hour and works 50 hours?

....

3. Marcie baby-sat for 30 hours over two weeks. She earned $1.00 an hour. What was her gross pay?

....

4. If Janet earns $180.00 after working 20 hours what is the hourly rate?

....

5. How much will Donald earn if he earns $5.00 per hour and works 12 hours?

....

1. If Sandra earns $72.00 after working 18 hours what is the hourly rate?

2. How much will Billy earn if he earns $5.00 per hour and works 34 hours?

3. Marin baby-sat for 33 hours over two weeks. She earned $9.00 an hour. What was her gross pay?

4. Janet baby-sat for 28 hours over two weeks. She earned $6.00 an hour. What was her gross pay?

5. How much will Paul earn if he earns $6.00 per hour and works nine hours?

1. If Ellen earns $40.00 after working 20 hours what is the hourly rate?

   ....

2. Jackie baby-sat for 19 hours over two weeks. She earned $2.00 an hour. What was her gross pay?

   ....

3. How much will David earn if he earns $2.00 per hour and works 28 hours?

   ....

4. Marin baby-sat for 28 hours over two weeks. She earned $9.00 an hour. What was her gross pay?

   ....

5. How much will Billy earn if he earns $3.00 per hour and works 33 hours?

   ....

1. How much will Jake earn if he earns $6.00 per hour and works seven hours?

    ....

2. If Sharon earns $32.00 after working eight hours what is the hourly rate?

    ....

3. Jackie baby-sat for 34 hours over two weeks. She earned $2.00 an hour. What was her gross pay?

    ....

4. If Sandra earns $111.00 after working 37 hours what is the hourly rate?

    ....

5. How much will Paul earn if he earns $8.00 per hour and works 48 hours?

    ....

1. How much will Billy earn if he earns $3.00 per hour and works 29 hours?

....

2. Michele baby-sat for 10 hours over two weeks. She earned $5.00 an hour. What was her gross pay?

....

3. If Sharon earns $196.00 after working 28 hours what is the hourly rate?

....

4. How much will Adam earn if he earns $7.00 per hour and works 45 hours?

....

5. If Jennifer earns $100.00 after working 20 hours what is the hourly rate?

....

ANSWERS

hot dog = $1.00	cola = $1.00
order of French-fries = $1.00	ice cream cone = $1.50
hamburger = $2.00	milk shake = $2.00
deluxe cheeseburger = $3.00	taco = $2.50

1. $28.50   What is the total cost of five milk shakes, three hot dogs, two ice cream cones, and five tacos?

2. $7.00   Brian wants to buy an order of French-fries and four ice cream cones. How much will he have to pay?

3. $21.50   If Adam wanted to buy three milk shakes, five ice cream cones, and four hamburgers, how much money would he need?

4. $17.00   What is the total cost of two hamburgers, two hot dogs, a cola, and five milk shakes?

5. $18.00   If Donald wanted to buy four hamburgers, an order of French-fries, four ice cream cones, and a deluxe cheeseburger, how much money would he need?

6. $8.00   If Michele buys two colas, what will her change be if she pays $10.00?

7. $25.50   What is the total cost of five orders of French-fries, three hot dogs, five milk shakes, and three tacos?

8. $6.50   Steven purchases five milk shakes, two hamburgers, and three ice cream cones. How much money will he get back if he pays $25.00?

hot dog = $1.00	cola = $1.00
order of French-fries = $1.00	ice cream cone = $1.50
hamburger = $2.50	milk shake = $2.50
deluxe cheeseburger = $3.00	taco = $2.00

1. $4.00   What is the total cost of four colas?

2. $23.50   Jake wants to buy a cola, five ice cream cones, and five deluxe cheeseburgers. How much will he have to pay?

3. $7.50   What is the total cost of three hamburgers?

4. $26.00   If Billy wanted to buy three hamburgers, five deluxe cheeseburgers, a milk shake, and an order of French-fries, how much would he have to pay?

5. $10.00   What is the total cost of four orders of French-fries and four ice cream cones?

6. $19.50   If Marin wanted to buy a hot dog, three hamburgers, two deluxe cheeseburgers, and five orders of French-fries, how much would she have to pay?

7. $4.00   Sharon wants to buy two tacos. How much will it cost her?

8. $4.00   Jennifer purchases three tacos and four milk shakes. If she had $20.00, how much money will she have left?

hot dog = $1.50	cola = $1.00
order of French-fries = $0.50	ice cream cone = $1.50
hamburger = $2.00	milk shake = $2.00
deluxe cheeseburger = $3.50	taco = $2.00

1. $17.50   Marcie wants to buy four milk shakes, two colas, and five ice cream cones. How much will it cost her?

2. $10.00   David purchases five milk shakes. How much money will he get back if he pays $20.00?

3. $18.50   What is the total cost of four orders of French-fries, four milk shakes, two hot dogs, a taco, and a deluxe cheeseburger?

4. $13.50   If Amy buys two deluxe cheeseburgers, an order of French-fries, two hot dogs, and three tacos, what will her change be if she pays $30.00?

5. $8.00   Sandra purchases four orders of French-fries. If she had $10.00, how much money will she have left?

6. $4.00   Jennifer purchases four milk shakes, three orders of French-fries, two hamburgers, and five ice cream cones. How much change will she get back from $25.00?

7. $8.00   What is the total cost of four orders of French-fries and three milk shakes?

8. $10.00   If Adam wanted to buy a deluxe cheeseburger, two tacos, an ice cream cone, and two orders of French-fries, how much would he have to pay?

hot dog = $1.50	cola = $1.00
order of French-fries = $1.00	ice cream cone = $1.00
hamburger = $2.50	milk shake = $2.00
deluxe cheeseburger = $3.00	taco = $2.50

1. $7.00   If Jennifer buys three ice cream cones, what will her change be if she pays $10.00?

2. $19.00   What is the total cost of five orders of French-fries, three hot dogs, three hamburgers, and a milk shake?

3. $5.50   If Ellen buys four deluxe cheeseburgers, two tacos, four ice cream cones, a hot dog, and a milk shake, and if she had $30.00, how much money will she have left?

4. $14.00   Audrey purchases three milk shakes. What will her change be if she pays $20.00?

5. $13.00   If Adam buys three deluxe cheeseburgers, two milk shakes, three hamburgers, five tacos, and four orders of French-fries, and if he had $50.00, how much money will he have left?

6. $4.00   What is the total cost of four ice cream cones?

7. $11.50   If Jake buys four hot dogs and a hamburger, how much money will he get back if he pays $20.00?

8. $29.00   Jackie wants to buy five milk shakes, four deluxe cheeseburgers, three colas, a hot dog, and a hamburger. How much will it cost her?

hot dog = $1.00	cola = $1.00
order of French-fries = $0.50	ice cream cone = $1.00
hamburger = $2.50	milk shake = $2.00
deluxe cheeseburger = $3.50	taco = $2.00

1. $12.50   If Billy wanted to buy an order of French-fries, four ice cream cones, and four tacos, how much would he have to pay?

2. $10.00   Amy wants to buy five milk shakes. How much will she have to pay?

3. $13.50   Brian purchases four ice cream cones and a hamburger. How much money will he get back if he pays $20.00?

4. $26.00   Allan wants to buy two milk shakes, five tacos, three hamburgers, a deluxe cheeseburger, and a hot dog. How much will it cost him?

5. $9.00   Jackie wants to buy five hot dogs and two tacos. How much will it cost her?

6. $11.50   Ellen purchases a deluxe cheeseburger and five ice cream cones. How much change will she get back from $20.00?

7. $7.00   If Donald buys five tacos, four hamburgers, and three ice cream cones, what will his change be if he pays $30.00?

8. $8.50   Steven purchases three orders of French-fries. How much money will he get back if he pays $10.00?

hot dog = $1.00	cola = $1.00
order of French-fries = $1.00	ice cream cone = $1.50
hamburger = $2.50	milk shake = $2.50
deluxe cheeseburger = $3.50	taco = $2.00

1. $6.50   Michele purchases three milk shakes, two orders of French-fries, and four colas. If she had $20.00, how much money will she have left?

2. $6.00   Allan wants to buy three tacos. How much will it cost him?

3. $1.00   What is the total cost of a hot dog?

4. $5.00   If Jake buys five orders of French-fries, how much change will he get back from $10.00?

5. $16.00   Amy wants to buy a milk shake, a cola, an order of French-fries, five ice cream cones, and four hot dogs. How much will she have to pay?

6. $7.00   If Sandra buys three colas and five tacos, what will her change be if she pays $20.00?

7. $9.00   What is the total cost of a hot dog and four tacos?

8. $11.50   If Jennifer buys three hamburgers, four tacos, and three orders of French-fries, how much money will she get back if she pays $30.00?

hot dog = $1.50	cola = $1.00
order of French-fries = $0.50	ice cream cone = $1.00
hamburger = $2.00	milk shake = $2.00
deluxe cheeseburger = $3.50	taco = $2.50

1. $21.50   What is the total cost of a milk shake, five tacos, and two deluxe cheeseburgers?

2. $9.50   Ellen wants to buy five ice cream cones, an order of French-fries, and two hamburgers. How much money will she need?

3. $36.50   If Audrey wanted to buy five tacos, four deluxe cheeseburgers, two colas, and four milk shakes, how much money would she need?

4. $5.50   If Michele buys three deluxe cheeseburgers and two hamburgers, how much money will she get back if she pays $20.00?

5. $14.50   If Sandra wanted to buy two hamburgers and three deluxe cheeseburgers, how much money would she need?

6. $16.50   If Donald wanted to buy three orders of French-fries, a hot dog, three tacos, two ice cream cones, and four colas, how much would it cost him?

7. $19.50   If Sharon wanted to buy an ice cream cone, four deluxe cheeseburgers, three orders of French-fries, and two hot dogs, how much money would she need?

8. $7.00   Billy purchases four tacos, three hamburgers, two orders of French-fries, two hot dogs, and three colas. How much change will he get back from $30.00?

hot dog = $1.50	cola = $1.00
order of French-fries = $1.00	ice cream cone = $1.00
hamburger = $2.00	milk shake = $2.00
deluxe cheeseburger = $3.00	taco = $2.00

1. $36.00   What is the total cost of five deluxe cheeseburgers, five tacos, three hamburgers, and five colas?

2. $3.00   Adam purchases two colas. What will his change be if he pays $5.00?

3. $3.50   What is the total cost of a hamburger and a hot dog?

4. $26.50   Sandra wants to buy a milk shake, five deluxe cheeseburgers, an ice cream cone, three hot dogs, and two tacos. How much will it cost her?

5. $9.00   Audrey wants to buy four ice cream cones and five colas. How much money will she need?

6. $7.00   If Sharon buys an ice cream cone and four deluxe cheeseburgers, how much change will she get back from $20.00?

7. $3.00   Jackie wants to buy a deluxe cheeseburger. How much will she have to pay?

8. $9.00   David purchases three tacos and five deluxe cheeseburgers. What will his change be if he pays $30.00?

hot dog = $1.00	cola = $1.00
order of French-fries = $0.50	ice cream cone = $1.50
hamburger = $2.50	milk shake = $2.50
deluxe cheeseburger = $3.50	taco = $2.50

1. $22.50 Steven wants to buy two colas, two deluxe cheeseburgers, a hot dog, and five milk shakes. How much will he have to pay?

2. $20.00 Janet wants to buy five hot dogs, three hamburgers, and three tacos. How much money will she need?

3. $11.00 If David buys three colas, three hamburgers, two orders of French-fries, two milk shakes, and a taco, how much money will he get back if he pays $30.00?

4. $13.50 Sharon purchases five orders of French-fries, an ice cream cone, and five tacos. How much money will she get back if she pays $30.00?

5. $10.00 Marin wants to buy two ice cream cones, two colas, and five hot dogs. How much will she have to pay?

6. $5.00 If Donald wanted to buy two tacos, how much would he have to pay?

7. $8.00 If Jackie wanted to buy a taco, two hamburgers, and an order of French-fries, how much would she have to pay?

8. $12.50 If Ellen buys four colas, a hot dog, four milk shakes, five orders of French-fries, and four hamburgers, what will her change be if she pays $40.00?

hot dog = $1.00	cola = $1.00
order of French-fries = $1.00	ice cream cone = $1.50
hamburger = $2.00	milk shake = $2.50
deluxe cheeseburger = $3.00	taco = $2.50

1. $4.00 If David buys two tacos and two deluxe cheeseburgers, and if he had $15.00, how much money will he have left?

2. $26.50 If Donald wanted to buy two ice cream cones, two hot dogs, four tacos, three milk shakes, and two hamburgers, how much money would he need?

3. $22.50 If Allan wanted to buy two tacos, four orders of French-fries, three ice cream cones, and three deluxe cheeseburgers, how much money would he need?

4. $2.00 What is the total cost of a hamburger?

5. $21.50 Audrey wants to buy a hamburger, two ice cream cones, five orders of French-fries, three deluxe cheeseburgers, and a taco. How much will she have to pay?

6. $5.00 Sandra wants to buy two milk shakes. How much will it cost her?

7. $14.00 Janet purchases three hamburgers. How much change will she get back from $20.00?

8. $11.00 Jake purchases two hamburgers, five tacos, five deluxe cheeseburgers, three hot dogs, and three ice cream cones. If he had $50.00, how much money will he have left?

1. = $45.50

2. = $35.50

3. = $11.50

4. = $12.00

5. = $41.00

6. = $41.00

7. = $11.00

8. = $51.00

1. = $26.00

2. = $50.00

3. = $16.00

4. = $45.50

5. = $26.50

6. = $6.00

7. = $31.00

8. = $65.00

Page 13   Emma.School

1.   = $20.00
        _____   _____   _____

2.   = $36.00
        _____   _____   _____   _____

3.   = $56.00
        _____   _____   _____   _____
        _____

4.   = $21.00
        _____   _____   _____   _____

5.   = $22.50
        _____   _____   _____

6.   = $31.00
        _____   _____   _____

7.   = $12.50
        _____   _____   _____

8.   = $42.00
        _____   _____   _____   _____
        _____

Page 14   Emma.School

1.   = $15.50
        _____   _____   _____

2.   = $35.50
        _____   _____   _____   _____   _____

3.   = $20.50
        _____   _____   _____

4.   = $41.00
        _____   _____   _____   _____
        _____

5.   = $20.50
        _____   _____   _____

6.   = $30.00
        _____   _____   _____   _____

7.   = $17.50
        _____   _____   _____   _____

8.   = $27.00
        _____   _____   _____

Page 15   Emma.School

1.   = $21.50
        _____   _____   _____   _____

2.   = $55.50
        _____   _____   _____   _____   _____

3.   = $21.50
        _____   _____   _____

4.   = $41.50
        _____   _____   _____   _____

5.   = $40.50
        _____   _____   _____

6.   = $12.00
        _____   _____   _____   _____

7.   = $30.00
        _____   _____   _____

8.   = $50.50
        _____   _____   _____   _____

Page 16   Emma.School

1.   = $40.50
        _____   _____   _____

2.   = $55.00
        _____   _____   _____   _____

3.   = $21.50
        _____   _____   _____   _____

4.   = $7.00
        _____   _____   _____   _____

5.   = $15.50
        _____   _____   _____

6.   = $46.50
        _____   _____   _____   _____   _____

7.   = $10.50
        _____   _____   _____

8.   = $12.50
        _____   _____   _____   _____

1. = $11.50

2. = $10.50

3. = $47.00

4. = $31.50

5. = $15.50

6. = $21.00

7. = $31.00

8. = $30.00

1. = $22.50

2. = $35.50

3. = $30.50

4. = $6.50

5. = $31.50

6. = $12.00

7. = $21.50

8. = $80.00

1. = $22.00

2. = $36.00

3. = $26.00

4. = $42.00

5. = $35.00

6. = $12.00

7. = $55.00

8. = $40.50

1. = $4.50

2. = $16.00

3. = $15.50

4. = $21.50

5. = $6.00

6. = $26.50

7. = $50.50

8. = $11.00

1. How much will David earn if he earns $4.00 per hour and works 23 hours?

    $92.00

2. If Marin earns $308.00 after working 44 hours what is the hourly rate?

    $7.00

3. Jennifer baby-sat for 30 hours over two weeks. She earned $9.00 an hour. What was her gross pay?

    $270.00

4. How much will Donald earn if he earns $4.00 per hour and works 48 hours?

    $192.00

5. If Ellen earns $225.00 after working 25 hours what is the hourly rate?

    $9.00

1. How much will Jake earn if he earns $7.00 per hour and works 32 hours?

    $224.00

2. If Michele earns $91.00 after working 13 hours what is the hourly rate?

    $7.00

3. Janet baby-sat for 49 hours over two weeks. She earned $4.00 an hour. What was her gross pay?

    $196.00

4. If Jennifer earns $82.00 after working 41 hours what is the hourly rate?

    $2.00

5. Marcie baby-sat for eight hours over two weeks. She earned $1.00 an hour. What was her gross pay?

    $8.00

1. If Marin earns $238.00 after working 34 hours what is the hourly rate?

    $7.00

2. How much will Adam earn if he earns $7.00 per hour and works 43 hours?

    $301.00

3. Ellen baby-sat for 24 hours over two weeks. She earned $3.00 an hour. What was her gross pay?

    $72.00

4. How much will Donald earn if he earns $3.00 per hour and works eight hours?

    $24.00

5. Jackie baby-sat for 47 hours over two weeks. She earned $1.00 an hour. What was her gross pay?

    $47.00

1. How much will Donald earn if he earns $4.00 per hour and works 23 hours?

    $92.00

2. Ellen baby-sat for 38 hours over two weeks. She earned $3.00 an hour. What was her gross pay?

    $114.00

3. If Jennifer earns $108.00 after working 36 hours what is the hourly rate?

    $3.00

4. Janet baby-sat for 36 hours over two weeks. She earned $9.00 an hour. What was her gross pay?

    $324.00

5. How much will David earn if he earns $9.00 per hour and works 44 hours?

    $396.00

1. If Janet earns $234.00 after working 39 hours what is the hourly rate?

    $6.00

2. Jennifer baby-sat for 28 hours over two weeks. She earned $9.00 an hour. What was her gross pay?

    $252.00

3. How much will Brian earn if he earns $3.00 per hour and works 38 hours?

    $114.00

4. If Marcie earns $11.00 after working 11 hours what is the hourly rate?

    $1.00

5. Audrey baby-sat for 25 hours over two weeks. She earned $3.00 an hour. What was her gross pay?

    $75.00

1. If Jennifer earns $45.00 after working 45 hours what is the hourly rate?

    $1.00

2. How much will Allan earn if he earns $3.00 per hour and works 50 hours?

    $150.00

3. Marcie baby-sat for 30 hours over two weeks. She earned $1.00 an hour. What was her gross pay?

    $30.00

4. If Janet earns $180.00 after working 20 hours what is the hourly rate?

    $9.00

5. How much will Donald earn if he earns $5.00 per hour and works 12 hours?

    $60.00

1. If Sandra earns $72.00 after working 18 hours what is the hourly rate?

    $4.00

2. How much will Billy earn if he earns $5.00 per hour and works 34 hours?

    $170.00

3. Marin baby-sat for 33 hours over two weeks. She earned $9.00 an hour. What was her gross pay?

    $297.00

4. Janet baby-sat for 28 hours over two weeks. She earned $6.00 an hour. What was her gross pay?

    $168.00

5. How much will Paul earn if he earns $6.00 per hour and works nine hours?

    $54.00

1. If Ellen earns $40.00 after working 20 hours what is the hourly rate?

    $2.00

2. Jackie baby-sat for 19 hours over two weeks. She earned $2.00 an hour. What was her gross pay?

    $38.00

3. How much will David earn if he earns $2.00 per hour and works 28 hours?

    $56.00

4. Marin baby-sat for 28 hours over two weeks. She earned $9.00 an hour. What was her gross pay?

    $252.00

5. How much will Billy earn if he earns $3.00 per hour and works 33 hours?

    $99.00

1. How much will Jake earn if he earns $6.00 per hour and works seven hours?

   $42.00

2. If Sharon earns $32.00 after working eight hours what is the hourly rate?

   $4.00

3. Jackie baby-sat for 34 hours over two weeks. She earned $2.00 an hour. What was her gross pay?

   $68.00

4. If Sandra earns $111.00 after working 37 hours what is the hourly rate?

   $3.00

5. How much will Paul earn if he earns $8.00 per hour and works 48 hours?

   $384.00

1. How much will Billy earn if he earns $3.00 per hour and works 29 hours?

   $87.00

2. Michele baby-sat for 10 hours over two weeks. She earned $5.00 an hour. What was her gross pay?

   $50.00

3. If Sharon earns $196.00 after working 28 hours what is the hourly rate?

   $7.00

4. How much will Adam earn if he earns $7.00 per hour and works 45 hours?

   $315.00

5. If Jennifer earns $100.00 after working 20 hours what is the hourly rate?

   $5.00

Printed in Great Britain
by Amazon

21047849R00032